BEGINNERS GUIDE TO GROWING AQUACULTURE FOR PROFIT

A Comprehensive Guide to Nurturing Thriving Aquaculture Gardens, Revealing Nature's Symphony in Your Backyard Oasis

Lucille Sarron

Table Of Contents

CHAPTER 1

Aquaculture Gardening

Aquaculture gardening emphasizes both economic and environmental sustainability and is a dynamic and sustainable method of raising aquatic creatures for commercial use. This creative farming technique has become well-known as an essential part of the world's food production system, making a substantial contribution to the world's seafood supply. The practice of aquaculture gardening entails the controlled environment of ponds, tanks, or underwater cages for the production of a variety of aquatic species, including fish, mollusks, crustaceans, and plants.

Just A Quick Rundown On Aquaculture Gardening

Over time, aquaculture gardening has changed to meet the expanding demand for seafood, relieve pressure on wild fish supplies, and provide a growing global population with a consistent supply of nutrition.

Aquaculture gardening makes it possible to manage the breeding, feeding, and harvesting of aquatic species, in contrast to conventional fishing, which is vulnerable to overfishing and environmental issues. This technique offers a more productive and ecological way to produce fish.

Pond aquaculture, recirculating aquaculture systems (RAS), and aquaponics are a few examples of aquaculture gardening systems. Pond aquaculture is the practice of raising fish or shrimp in man-made or natural ponds, while RAS uses water recirculation and sophisticated filtration to maximize the growing conditions for fish. By combining hydroponic gardening with aquaculture, aquaponics fosters a symbiotic interaction between fish and plants.

Aquaculture Gardening Is Important For Profits:

1. Stability and Predictability: When compared to conventional fishing, aquaculture gardening provides a more steady and predictable production cycle. Farmers have more control over environmental variables like temperature, feed, and water quality, which leads to regular yields and less reliance on uncontrollable elements like weather and ocean conditions.

2. Enhanced Productivity: Compared to conventional fisheries, aquaculture gardening offers increased output rates by creating ideal growing conditions. Because of this higher productivity, there is a more consistent and plentiful supply of seafood available to fulfill the market's increasing demand.

3. Diversification of Species: A variety of aquatic species, including fish, shellfish, and plants, can be cultivated through aquaculture gardening. Farmers

may investigate niche markets for certain species and adjust to market needs thanks to this variety, which boosts profitability.

4. Decreased Stress on Wild Fisheries: With the world's seafood demand only growing, aquaculture gardening is essential to easing the strain on wild fish populations. Sustainable aquaculture methods support biodiversity preservation and the preservation of marine ecosystems.

5. Employment Creation and Economic Impact: The expansion of aquaculture gardening not only offers chances for both small and large-scale farmers, but also boosts employment in associated sectors like processing, feed manufacturing, and equipment manufacturing. Local and regional communities benefit economically as a result of this.

6. Technological Developments: Continued research and technological developments in aquaculture gardening have produced

breakthroughs in disease control, feed formulation, and effective water use. Aquaculture farmers benefit from lower costs, more efficiency, and higher profitability as a result of these developments.

To sum up, aquaculture gardening is an essential part of today's agricultural landscape and offers a financially sound way to satisfy the growing demand for seafood around the world. It provides a dependable and controlled method of aquatic production that supports food security, economic growth, and environmental sustainability, making it significantly more profitable than traditional fisheries.

CHAPTER 2

Comprehending Aquaculture Gardening

Definition And Scope: Hydroponics, or soilless plant growth, and aquaculture, or the cultivation of aquatic organisms, are combined in aquaculture gardening, also referred to as aquaponics, a sustainable and integrated farming system. This novel method fosters a symbiotic relationship between fish and plants in which the plants filter and purify the fish's water and the fish waste supplies vital nutrients for plant growth. Aquaculture gardening is a closed-loop system that maximizes resource efficiency and minimizes waste.

By utilizing the mutual benefits that fish and plants have, aquaculture gardening goes beyond conventional farming. This technique has the potential to produce crops and fish in a regulated setting, encourage resource conservation, lessen the negative effects on the environment, and offer a reliable supply of food.

Historical Context: Although the idea of aquaculture gardening dates back to traditional agricultural methods, its current incarnation first appeared in the 20th century. The need for more effective and sustainable food production techniques led to the rise in popularity of the combination of hydroponics and aquaculture in the 1970s and 1980s. In order to maximize environmental impact while maximizing the use of water, nutrients, and space, researchers and practitioners looked for solutions.

The development of aquaculture gardening systems was greatly aided by the innovative work of aquaponics enthusiasts and researchers like Drs. James Rakocy and Mark McMurtry. These systems have been improved through experimentation and technological advancements over time, opening them up to a wider audience.

Types Of Aquaculture Systems: There are several types of aquaculture gardening systems, each with special characteristics and advantages. Among the well-known varieties are:

1. Media-Based Aquaponics: In this system, water is pumped through a soilless medium, like gravel or clay pellets, to support plant growth. In addition to supporting plant roots, the media serves as a biofilter, eliminating extraneous nutrients.

2. The Nutrient Film Technique (NFT) allows for efficient nutrient absorption by covering plant roots with a thin layer of nutrient-rich water. NFT systems are renowned for being easy to use and adaptable to a variety of crops.

3. Deep Water Culture (DWC): In DWC systems, plants are suspended with their roots submerged in nutrient-rich water. In these systems, oxygenation is

essential for maintaining the health of the fish and the plants.

4. Vertical aquaponics: By making use of vertical space, this design raises the plant density. Systems for vertical aquaponics are suited for urban farming and require little space.

5. Integrated Aquaculture: In order to create a more resilient and diversified farming ecosystem, some systems combine aquaculture with other farming components, such as livestock or poultry.

It is crucial for both practitioners and enthusiasts to comprehend the various forms of aquaculture gardening as well as its historical background. As technology develops, aquaculture gardening has the potential to be a major solution to the problems associated with global food security because it provides a resource-efficient and sustainable means of producing food.

CHAPTER 3

Organizing Your Aquaculture Garden

Growing aquatic life for food or other valuable resources, such as fish, shellfish, and aquatic plants, is known as aquaculture gardening. For your aquaculture endeavor to be successful and sustainable, careful planning is essential. Three main facets of planning will be discussed here: the selection of the site, environmental factors, and legal and regulatory obligations.

Site Choosing:

1. Water Purity:

• Since water quality has an immediate effect on the growth and health of aquatic organisms, pick a location with high water quality.

• Check that the water's pH, temperature, dissolved oxygen content, and nutrient levels satisfy the needs of the selected species.

2. Availability:

• Make sure it's simple to maintain, harvest, and monitor the location.

• Easy access to transportation hubs and markets is also necessary for the effective distribution of your aquaculture goods.

3. Facilities:

• Evaluate the state of the infrastructure, including the availability of waste disposal, water, and electricity.

• Enough room should be taken into account for building ponds, tanks, or other aquaculture structures.

Environmental Factors To Be Considered:

1. Impact on Ecosystem:

• Assess the possible effects of aquaculture on the regional ecology. Steer clear of areas that house endangered or delicate ecosystems.

• Use best management practices to reduce the negative effects on the environment, such as runoff of nutrients and effluent discharge.

2. Weather and Climate:

• Take into account the local weather and climate. Particular temperature ranges are favorable to some species, and aquaculture operations may be impacted by severe weather.

• To lessen the effects of bad weather, put up protective structures or make use of the right technology.

3. Conservation of Biodiversity:

• Include conservation strategies for biodiversity in your aquaculture plan. Steer clear of invasive species and adopt methods that enhance biodiversity in the area.

Regulation And Law Requirements:

1. Granting

• Look into and acquire the licenses and permits your aquaculture business needs. Licenses for aquaculture facilities, environmental impact assessments, and water use permits are a few examples of this.

2. Zoning laws:

• Make sure your aquaculture site is properly designated for such activities by adhering to local zoning regulations. Aquaculture operations' permitted land use and intensity may be governed by zoning laws.

3. Adherence to Environmental Standards:

• Comply with environmental laws to stop pollution and preserve the quality of the water. Put plans in place to reduce the amount of chemicals and effluents that are released into the nearby ecosystems.

4. Species Selection:

• Find out if the species you want to grow is subject to any restrictions. In order to prevent ecological disruptions, certain regions have regulations pertaining to the introduction of non-native species.

5. Participation in the Community:

• Interact with stakeholders and the local community to address issues and win support. Developing good relationships will improve your aquaculture project's social sustainability.

In conclusion, the success of your aquaculture gardening endeavor depends on meticulous planning that takes into account site selection, environmental considerations, and legal requirements. You support the long-term sustainability of your business and the well-being of the local ecosystem by placing a high priority on sustainability and compliance.

CHAPTER 4

Infrastructure And Necessary Equipment

Aquaculture gardening requires careful planning and investment in essential equipment and infrastructure to ensure the optimal growth and health of aquatic organisms. Here are key concepts related to tanks and ponds, aeration systems, and water filtration and recirculation:

1. Tanks And Ponds:

• Selection and Design: The choice of tanks or ponds depends on the type of aquaculture and the species being cultivated. Tanks can be above-ground or in-ground, while ponds may be earthen or lined. The design should consider factors like water volume, shape, and ease of maintenance.

• Material: Tanks and pond liners are typically made of materials like fiberglass, concrete, or high-density polyethylene. The material should be durable, non-toxic, and resistant to corrosion to ensure a long lifespan.

• Water Management: Proper water management is crucial. Tanks and ponds should have efficient inflow and outflow systems to maintain water quality and circulation. Regular monitoring of water parameters such as temperature, pH, and dissolved oxygen is essential.

2. Aeration Systems:

• Importance of Aeration: Aeration is vital for maintaining optimal oxygen levels in the water. It helps prevent oxygen depletion, promotes aerobic bacteria activity, and enhances the overall health of aquatic organisms.

• Types of Aeration Systems: There are various aeration systems, including diffused aeration, surface aerators, and paddlewheel aerators. The choice depends on factors like pond depth, size, and the oxygen requirements of the cultured species.

• Energy Efficiency: Sustainable aquaculture practices often prioritize energy-efficient aeration systems. Solar-powered aerators and low-energy consumption devices can reduce operational costs and minimize the environmental impact.

3. Water Filtration And Recirculation:

• Filtration Methods: Efficient water filtration is essential for removing debris, uneaten feed, and waste products. Mechanical filters, biological filters, and chemical filters are commonly used. Biofilters, in particular, promote the growth of beneficial bacteria that break down harmful substances.

• Recirculation Systems: Recirculating aquaculture systems (RAS) are designed to reuse and treat water, minimizing the need for constant water exchange. This approach is environmentally sustainable and can be particularly beneficial in areas with limited water resources.

• Monitoring and Control: Automation and monitoring systems play a key role in water filtration and recirculation. Sensors and control mechanisms help maintain optimal water quality parameters, ensuring a stable and healthy environment for aquaculture.

In summary, investing in the right tanks and ponds, effective aeration systems, and robust water filtration and recirculation infrastructure is crucial for the success of aquaculture gardening. These elements contribute to the overall sustainability, productivity, and well-being of the aquatic ecosystem in an aquaculture setting.

CHAPTER 5

Choosing The Right Fish And Plants

1. Profitable Fish Species: When establishing an aquaculture garden, selecting the right fish species is crucial for both ecological balance and financial success. Consider species that are in demand in the market and have a rapid growth rate. Popular choices include tilapia, catfish, trout, and carp. Conduct thorough market research to identify the preferences of local consumers and restaurants. Additionally, take into account the environmental conditions of your aquaculture system, such as water temperature and quality, to ensure the chosen fish species can thrive.

2. Compatible Plant Varieties: Integrating plants into your aquaculture system not only enhances the aesthetics but also contributes to the overall health of the ecosystem. Choose plant varieties that are compatible with the specific needs of your fish species. Some common choices include watercress, water spinach, and various types of lettuce. Plants can help maintain water quality by absorbing excess nutrients, providing shade, and acting as natural filters. Consider the growth habits of the plants to ensure they do not interfere with the swimming patterns or feeding behavior of the fish.

3. Symbiotic Relationships In Aquaculture:

Cultivating symbiotic relationships between fish and plants can create a balanced and self-sustaining ecosystem. In aquaponics, for example, fish waste provides essential nutrients for plants, while the plants filter and purify the water for the fish. This mutually beneficial relationship fosters a closed-loop system where the by-products of one component become inputs for another. Research and implement symbiotic combinations that suit your specific aquaculture system. For instance, some fish species thrive on the detritus produced by certain plants, creating a natural cycle of nutrient recycling.

4. Consideration Of Environmental Factors: Take into account the environmental conditions of your location when selecting fish and plants. Factors such as water temperature, pH levels, and sunlight exposure play a significant role in the success of your aquaculture garden. Ensure that the chosen fish and plants can withstand the prevailing environmental conditions to minimize the risk of disease outbreaks and poor growth. Additionally, consider the availability of resources like fresh water and feed, as these are critical for sustaining a healthy and productive aquaculture system.

5. Diversification For Stability: Diversifying your aquaculture garden by incorporating a variety of fish and plant species can contribute to system stability. Different species may have varying nutrient requirements, growth rates, and tolerances to environmental conditions. This diversity can enhance the resilience of your aquaculture system, reducing the risk of catastrophic failures and optimizing overall productivity.

In conclusion, the success of an aquaculture garden hinges on the careful selection of fish and plants, taking into consideration their compatibility, profitability, and environmental requirements. By fostering symbiotic relationships and diversifying your system, you can create a sustainable and thriving aquaculture garden that benefits both the ecosystem and your bottom line.

CHAPTER 6

Water Quality Management

Aquaculture gardening involves the cultivation of aquatic organisms, such as fish, shellfish, and plants, in controlled environments. One crucial aspect of successful aquaculture is water quality management. Maintaining optimal water conditions is essential for the health and growth of aquatic species. In this context, monitoring parameters like pH, temperature, and oxygen levels, as well as managing algae and microorganisms, play pivotal roles.

Monitoring Parameters:

1. PH Levels:

• Importance: pH measures the acidity or alkalinity of water. Most aquatic organisms thrive in a specific pH range, and deviations can stress or harm them.

• Monitoring and Control: Regularly measure pH levels using a reliable pH meter. Adjustments can be made using pH buffers or chemical treatments to maintain the desired range.

2. Warmth:

• Importance: Temperature affects metabolic rates, growth, and reproduction in aquatic organisms. Different species have specific temperature requirements.

• Monitoring and Control: Use temperature sensors to monitor water temperature. Employ heaters or coolers to regulate temperature, ensuring it stays within the optimal range for the targeted species.

3. Oxygen Levels:

• Importance: Adequate oxygen is crucial for the survival of aquatic organisms. Insufficient oxygen can lead to stress, disease, or mortality.

• Monitoring and Control: Use dissolved oxygen meters to measure oxygen levels. Aeration systems, such as diffusers or surface agitators, can be employed to enhance oxygen exchange.

Managing Algae And Microorganisms:

1. Algae Control:

• Importance: Algae can compete for nutrients, reduce oxygen levels at night, and create unfavorable conditions for aquaculture species.

• Preventive Measures: Implement shading techniques, use nutrient control strategies, and introduce algae-eating organisms like certain fish or invertebrates.

2. Microorganism Management:

• Importance: Harmful microorganisms can cause diseases and impact the health of aquatic organisms.

• Biosecurity Measures: Quarantine new stock, maintain good hygiene practices, and use probiotics to promote beneficial microbial communities. Regular water testing can help identify and address microbial issues promptly.

Overall Best Practices:

1. Regular Monitoring:

• Conduct routine checks for all parameters to identify trends or sudden changes.

• Keep accurate records of monitoring data for analysis and trend identification.

2. Adaptive Management:

• Adjust management practices based on seasonal changes, species requirements, and monitoring results.

3. Education and Training:

• Ensure that personnel involved in aquaculture understand the importance of water quality management and are trained in monitoring techniques.

4. Technology Integration:

• Utilize advanced technologies such as automated monitoring systems and data analytics to enhance efficiency and accuracy.

By adopting a comprehensive water quality management approach that includes monitoring and managing key parameters, aquaculture practitioners can create and maintain an optimal environment for the successful cultivation of aquatic species. This approach promotes the overall health, growth, and sustainability of the aquaculture system.

CHAPTER 7

Feeding And Nutrition

In aquaculture gardening, the concept of feeding and nutrition plays a pivotal role in ensuring the health and growth of both fish and plants. Proper nutrition is essential for achieving optimal yields, maintaining water quality, and promoting the overall sustainability of the system.

Proper Nutrition For Fish And Plants:

Fish and plants in aquaculture systems require a well-balanced and species-specific diet to thrive. Just like any other living organism, they need essential nutrients such as proteins, carbohydrates, fats, vitamins, and minerals. The nutritional requirements vary among different species, life stages, and environmental conditions.

For fish, the protein content of the diet is crucial for growth and muscle development. Carbohydrates provide energy, while fats serve as a concentrated energy source. Additionally, vitamins and minerals are essential for metabolic processes, reproduction, and overall well-being. Plants, on the other hand, require nutrients like nitrogen, phosphorus, potassium, and micronutrients for proper growth and development.

Achieving proper nutrition involves understanding the specific dietary needs of the aquatic organisms in the system. Commercially available feeds are often formulated to meet these requirements, and their usage should be optimized based on the growth stage and size of the fish or the growth phase of the plants.

Sustainable Feed Options:

In the context of aquaculture gardening, sustainability is a key consideration. Traditional fish feeds often rely on marine resources, such as fishmeal and fish oil, which can contribute to overfishing and environmental degradation. Sustainable feed options aim to address these concerns by incorporating alternative protein and lipid sources.

Alternative protein sources for fish feed include plant proteins, insect meal, and single-cell proteins derived from microorganisms. These alternatives not only reduce the reliance on wild-caught fish for feed but also contribute to a more sustainable and environmentally friendly aquaculture industry.

Feed Conversion Ratios:

Feed conversion ratio (FCR) is a crucial metric in aquaculture that measures the efficiency of converting feed into the edible biomass of the fish. A lower FCR indicates better efficiency, as less feed is required to produce a given amount of fish biomass. Monitoring and optimizing FCR can lead to reduced production costs, minimized environmental impact, and increased profitability.

Efforts to improve FCR often involve the use of high-quality feeds, optimizing feeding practices, and selecting fish species that exhibit efficient growth characteristics. Sustainable aquaculture practices emphasize the importance of achieving lower FCRs to promote resource efficiency and reduce the ecological footprint of aquaculture operations.

In conclusion, feeding and nutrition are fundamental aspects of aquaculture gardening. By focusing on proper nutrition, exploring sustainable feed options, and optimizing feed conversion ratios, practitioners can foster a balanced and environmentally conscious approach to aquaculture that benefits both fish and plants within the system.

CHAPTER 8

Disease Prevention And Health Management

Aquaculture gardening involves the cultivation of aquatic organisms such as fish, shellfish, and plants in controlled environments. One of the key challenges faced by aquaculturists is the management of diseases, which can significantly impact the health and productivity of the aquatic ecosystem. Implementing effective disease prevention and health management strategies is crucial to ensuring the success and sustainability of aquaculture operations.

Common Aquaculture Diseases:

1. Bacterial Infections:

• Aeromoniasis and Motile Aeromonad Septicemia (MAS) in fish.

• Vibriosis in marine aquaculture.

2. Viral Infections:

• Viral Nervous Necrosis (VNN) affecting various species.

• Infectious Pancreatic Necrosis (IPN) in salmonids.

3. Parasitic Infections:

• Ichthyophthirius multifiliis (Ich) causing white spot disease.

• Argulus spp. and Lernaea spp. as ectoparasites.

4. Fungal Infections:

• Saprolegnia spp. leading to saprolegniasis.

Biosecurity Measures: Effective biosecurity measures play a crucial role in preventing the introduction and spread of diseases in aquaculture systems. Key practices include:

1. Quarantine Protocols:

• Implementing quarantine periods for new stock to observe and detect any signs of diseases before introducing them to existing populations.

2. Site Selection and Design:

• Choosing sites with low disease risk and proper water quality.

• Designing facilities to minimize the risk of external contamination.

3. Water Management:

• Regular monitoring and maintenance of water quality parameters.

• Avoiding the use of water from questionable sources.

4. Stocking Density Management:

• Maintaining appropriate stocking densities to reduce stress and susceptibility to diseases.

5. Equipment and Personnel Hygiene:

• Disinfecting equipment regularly.

• Implementing strict hygiene protocols for personnel to prevent cross-contamination.

Medications And Treatments:

1. Prophylactic Measures:

• Use of vaccines to prevent the onset of specific diseases.

• Application of probiotics to enhance the immune system of aquatic organisms.

2. Chemotherapeutics:

• Application of antibiotics and antimicrobial agents to treat bacterial infections.

• Antiparasitic medications to control parasitic infestations.

3. Water Treatment:

• Application of disinfectants to treat waterborne pathogens.

• UV and ozone treatment to reduce the microbial load in recirculating aquaculture systems.

4. Nutritional Management:

• Providing a balanced and nutritionally rich diet to enhance the overall health and immune response of aquatic organisms.

Disease prevention and health management are integral components of successful aquaculture gardening. By implementing stringent biosecurity measures and employing appropriate medications and treatments, aquaculturists can mitigate the impact of diseases, ensuring the sustainable growth of their aquatic organisms and the overall success of their operations. Continuous research and development in this field are essential to stay ahead of emerging diseases and evolving challenges in aquaculture.

CHAPTER 9

Harvesting Techniques

Aquaculture gardening involves the cultivation of aquatic organisms in controlled environments for food production. Effective harvesting techniques are crucial to ensure a successful aquaculture operation. Here are key concepts related to harvesting in aquaculture gardening:

1. Optimal Harvest Times:

• Life Stage Monitoring: Regular monitoring of the life stages of the aquatic organisms is essential. Harvesting at the optimal time ensures maximum growth, yield, and quality.

• Environmental Factors: Consideration of environmental conditions such as water temperature, dissolved oxygen levels, and nutrient availability is vital. Harvesting during favorable conditions can minimize stress on the organisms.

2. Harvesting Methods:

• Netting and Seining: Commonly used for fish and shellfish, netting and seining involve the use of nets to corral and capture organisms. This method is effective for schools of fish or densely populated areas.

• Trawling: Suitable for larger species, trawling involves dragging a net through the water. It's efficient for harvesting organisms that swim near the surface.

• Dredging: Used for bivalve mollusks like oysters and clams, dredging involves collecting organisms from the substrate using a dredge.

• Grading and Sorting: In cases where different sizes are harvested, grading and sorting help separate organisms based on size, ensuring uniformity.

3. Post-Harvest Handling And Processing:

• Live Harvests: Some organisms, especially those destined for live markets, require careful handling to maintain their vitality. Proper packing and transportation methods are crucial to minimize stress and mortality.

• Processing Facilities: For species that undergo processing, having well-equipped facilities for cleaning, filleting, and packaging is essential. This step contributes to the quality and marketability of the product.

• Preservation Techniques: Immediate preservation methods such as chilling, freezing, or canning can extend the shelf life of harvested products. These techniques help maintain product quality during transportation and storage.

4. Quality Control:

• Monitoring Health: Regular health checks of the organisms before harvesting can help identify any signs of diseases or abnormalities. This ensures that only healthy organisms are harvested, contributing to the overall quality of the product.

• Water Quality: Ensuring the water quality during and after harvesting is crucial. Poor water quality can negatively impact the organisms and the harvested product. Adequate measures should be taken to maintain water quality throughout the process.

5. Environmental Sustainability:

• Selective Harvesting: Implementing selective harvesting practices can help minimize the impact on non-target species and reduce the ecological footprint of aquaculture operations.

• Efficient Resource Use: Optimize resource utilization during harvesting to minimize waste. Proper planning and technology adoption can contribute to sustainable aquaculture practices.

In conclusion, successful aquaculture harvesting involves a combination of strategic planning, monitoring, and efficient techniques. Considering optimal harvest times, employing suitable harvesting methods, implementing proper post-harvest handling and processing, maintaining quality control, and prioritizing environmental sustainability are key components in ensuring a thriving and sustainable aquaculture gardening operation.

CHAPTER 10

Marketing And Selling Your Aquaculture Products

Marketing and selling your aquaculture products is a crucial aspect of running a successful aquaculture gardening business. Effective marketing strategies can help you reach potential customers, create demand for your products, and ultimately boost sales. Here are key concepts related to marketing and selling in aquaculture gardening:

Identifying Target Markets

1. Market Research: Conduct thorough market research to understand the demand for your aquaculture products. Identify potential customers, their preferences, and the factors that influence their purchasing decisions. This information will help you

tailor your marketing efforts to meet specific market needs.

2. Segmentation: Divide your target market into segments based on demographics, geographic location, and psychographics. This segmentation allows you to create targeted marketing messages for each group, increasing the relevance of your products to different customer segments.

3. Consumer Trends: Stay updated on consumer trends in the aquaculture and seafood industry. This includes preferences for sustainably sourced products, organic options, and health-conscious choices. Adapting to these trends can give your products a competitive edge in the market.

Branding And Packaging

1. Brand Identity: Develop a strong brand identity for your aquaculture products. This includes creating a memorable logo, defining your brand values, and

establishing a unique selling proposition. A strong brand can differentiate your products from competitors and build customer loyalty.

2. Packaging Design: Invest in attractive and functional packaging. The packaging should not only protect the freshness and quality of your products but also stand out on the shelves. Consider eco-friendly packaging options

CHAPTER 11

Maximizing Profits And Sustainability

Aquaculture gardening involves the cultivation of aquatic organisms such as fish, shellfish, and plants in controlled environments. To ensure the long-term success of aquaculture operations, it is essential to focus on maximizing profits while embracing sustainable practices. This involves efficient cost management, scaling up operations, and incorporating sustainable principles into aquaculture processes.

1. Cost Management:

Effective cost management is crucial for the economic viability of aquaculture gardening. Consider the following strategies to optimize costs:

a. Feed Efficiency: Efficient feeding practices, including the use of high-quality feed and minimizing waste, can contribute to significant cost savings.

b. Energy Optimization: Implement energy-efficient technologies and practices to reduce operational costs. This includes efficient water circulation systems, lighting, and temperature control.

c. Disease Prevention: Invest in biosecurity measures to prevent disease outbreaks, as disease management can be a major cost in aquaculture. Regular health monitoring and timely interventions can minimize losses.

d. Economies of Scale: As operations expand, take advantage of economies of scale. Bulk purchasing, efficient labor allocation, and streamlined processes can lower production costs.

2. Scaling Up Operations:

Scaling up aquaculture operations is essential for maximizing profits. However, it should be done thoughtfully to ensure the sustainability of the venture:

a. Infrastructure Development: Invest in infrastructure that can accommodate increased production without compromising environmental sustainability. This includes upgrading tanks, water treatment facilities, and other critical components.

b. Market Research: Before scaling up, conduct thorough market research to ensure there is sufficient demand for increased production. Understanding market trends and consumer preferences is crucial for successful expansion.

c. Technology Adoption: Embrace technological advancements such as automation, data analytics, and remote monitoring to enhance efficiency as operations scale up. These technologies can optimize resource utilization and reduce labor costs.

3. Sustainable Practices In Aquaculture:

Sustainability is paramount in aquaculture gardening to ensure the long-term health of ecosystems and communities. Adopt the following sustainable practices:

a. Water Conservation: Implement water recycling systems to minimize water usage and reduce the environmental impact of aquaculture operations.

b. Certification Programs: Seek certification from recognized sustainable aquaculture programs, such as the Aquaculture Stewardship Council (ASC) or Best Aquaculture Practices (BAP), to demonstrate commitment to sustainable practices.

c. Biodiversity Conservation: Integrate biodiversity-friendly practices, such as protecting natural habitats and avoiding the use of harmful chemicals, to preserve the ecological balance in and around aquaculture facilities.

d. Community Engagement: Foster positive relationships with local communities by being transparent about operations, addressing concerns, and contributing to local development initiatives.

Conclusion:

Maximizing profits and sustainability in aquaculture gardening requires a holistic approach that combines cost management, strategic scaling of operations, and a commitment to sustainable practices. By balancing economic considerations with environmental and social responsibility, aquaculture ventures can thrive in the long term, meeting the growing demand for aquatic products while safeguarding the health of ecosystems and communities.

Summary

In conclusion, our exploration of aquaculture gardening has illuminated several key learnings that highlight the immense potential this field holds for sustainable agriculture, environmental conservation, and economic prosperity. As we recap these key insights, it becomes evident that aquaculture gardening is not merely a method of cultivating aquatic organisms; rather, it is a holistic approach that integrates aquaculture with gardening practices to create a symbiotic and efficient system.

One of the fundamental learnings is the importance of maintaining a balanced ecosystem within aquaculture gardens. The coexistence of fish and plants in a controlled environment fosters a mutually beneficial relationship. Fish waste provides essential nutrients for plant growth, while plants help filter and purify the water, creating an environment conducive to the well-being of both components.

Moreover, the potential for nutrient-rich yields from aquaculture gardening is a significant advantage. The combination of fish and plant cultivation allows for a diverse range of produce, providing a sustainable source of protein and fresh vegetables. This not only addresses food security concerns but also contributes to a healthier and more varied diet for communities engaged in aquaculture gardening.

As we encourage continued learning and improvement in aquaculture gardening, it is crucial to emphasize the role of technology and innovation. Constantly evolving techniques, such as aquaponics and hydroponics, enable us to optimize resource utilization and enhance productivity. Additionally, ongoing research and development in breeding programs can lead to the cultivation of more resilient and productive fish and plant varieties, further strengthening the sustainability of aquaculture gardening systems.

Acknowledging the potential for profits in aquaculture gardening is essential for promoting its widespread adoption. Beyond meeting local food demands, aquaculture gardening offers a lucrative opportunity for entrepreneurs and farmers alike. The diversification of income streams through the sale of fish, vegetables, and herbs can contribute to the economic stability of individuals and communities involved in this practice.

In conclusion, the future of aquaculture gardening is promising, with the potential to address global challenges related to food production, environmental sustainability, and economic development. By fostering a deeper understanding of the interconnectedness of aquatic and plant life, embracing technological advancements, and recognizing the financial prospects it presents, we can unlock the full potential of aquaculture gardening for the benefit of present and future generations.

As we move forward, let us remain committed to learning, adapting, and innovating to harness the full potential of this sustainable and profitable agricultural approach.

THE END

www.ingramcontent.com/pod-product-compliance
Lightning Source LLC
Chambersburg PA
CBHW070044260726
48658CB00002B/728